1ST EDITION

HOW TO BUILD A

SWIMMING POOL AT HOME

FEBRUARY 2023

11. POOL SAFETY: THIS CHAPTER ADDRESSES SAFETY ISSUES, INCLUDING FENCES, ALARMS, SAFETY DEVICES, CHILD SUPERVISION AND RULES FOR POOL USE.

12. CONCLUSION: CONCLUDE THE E-BOOK WITH A SUMMARY OF THE TOPICS COVERED AND REINFORCE THE IMPORTANCE OF PLANNING, CAREFUL INSTALLATION AND REGULAR MAINTENANCE OF THE SWIMMING POOL TO ENSURE ITS SAFE AND ENJOYABLE USE.

INTRODUCTION

Welcome to your complete guide to building a swimming pool at home! With this e-book, you will have all the information you need to plan and build the pool of your dreams.

Before we get started, it's important to consider why you might want a pool at home. Is it for recreational purposes, to improve your health, or simply to increase the value of your home?

Regardless of the reason, building a swimming pool at home is a major undertaking and requires careful planning and attention to detail.

In this ebook, we'll discuss the different types of pools, including cement, fiberglass, and natural pond pools, as well as the advantages and disadvantages of each option. In addition, we will cover important issues such as location, size, depth, shape and necessary equipment.

You will also learn about the construction process, from project design to equipment installation, and how to keep your pool in good condition for safe and enjoyable use.

This e-book is designed to be a practical, easy-to-follow guide to building a swimming pool at home, regardless of your level of knowledge on the subject. So let's start!

CHAPTER 2
POOL PLANNING

Before you start building your pool at home, it's important to take the time for proper planning. This will help ensure that your pool is built correctly, is safe and meets your needs and expectations. Here are some tips to help you plan your pool.

1. Choice of pool type: There are several types of pools available, including cement, fiberglass and natural lake pools. Each type has its own advantages and disadvantages, so it's important to carefully consider which type is best for your needs.

2. Choosing a location: The location of your pool is one of the most important decisions you will need to make. Consider sun and shade, wind, privacy, safety and cost before choosing your pool location.

3. Size and shape: Decide on the size and shape of your pool. This will depend on the space available, the use you want to make of the pool and your personal preferences.

4. Depth: Choose the desired depth for your pool. This could be a safety issue, especially if you have children, or a style issue.

5. Deck design: If you want to add a decking around your pool, it's important to plan ahead. Consider the height, material, color, and texture of the decking to match the pool and home.

6. Equipment: Decide what equipment you will need for your pool, such as pumps, filters, heaters and lighting systems.

7. Budget: Finally, it is important to establish a budget for your pool and include all costs including materials, labor, equipment and possible additional expenses.

Remember that careful, detailed planning is critical to ensuring your pool build is a success. The more time and effort you put into planning now, the more satisfied you'll be with the end result.

CHAPTER 3
TYPES OF SWIMMING

There are several types of pools available to choose from, including cement, fiberglass and natural pond pools. Here is a detailed description of each type, along with its advantages and disadvantages.

1. Cement pools: Cement pools are the most traditional and solid ones available. They are made of cement, mortar and ceramic or vinyl coating. Advantages include durability, versatility in terms of shape and size, and ease of maintenance. Disadvantages include high initial cost, longer construction time and the need for periodic maintenance.

2. Fiberglass Pools: Fiberglass pools are made of glass or fiberglass composite material. They are lightweight, easy to install and cost effective. Advantages include low cost, ease of installation and maintenance, and good looks. Disadvantages include the risk of scratches or cracks, the need for periodic maintenance, and limitations in terms of shape and size.

3. Natural Lake Pools: Natural lake pools are pools that mimic the appearance and ecosystem of a natural lake. They are made from rock, earth and plants, and require less chemistry and maintenance than other types of pools.

Advantages include natural appearance, low maintenance, low maintenance cost and less environmental impact. Disadvantages include the need for more space, the limitation in terms of size and shape, and the need for periodic maintenance.

When choosing the type of pool for your home, carefully consider the advantages and disadvantages of each type, as well as your needs, budget, and available space. By making an informed choice, you can be confident that your pool will be a great addition to your home for years to come.

CHAPTER 4
PREPARING THE GROUND

Before starting the construction of the pool, it is important to prepare the land properly. Here are the steps to ensure a solid foundation for your pool:

1. Check Local Regulations: Check for local regulations that may affect pool construction, such as building regulations, safety laws, and zoning restrictions. Make sure you get all the necessary permits and approvals before starting construction.

2. Choice of location: Choose the right location for your pool. Make sure there is enough space for the pool and that it will be located on a flat, level area. Also, make sure the terrain is clear of obstacles such as tree roots, rocks, or groundwater.

3. Obstacle Removal: Remove any obstacles present on the terrain, such as trees, bushes or rocks. Make sure the ground is level and clear before starting pool construction.

4. Soil compaction: Make sure the soil is well compacted before you start building the pool. This will ensure a solid foundation for your pool and prevent it from sinking or becoming unstable over time.

5. Installation of drains: If necessary, install drains to ensure that rainwater does not accumulate around the pool. This will also help protect the pool structure from damage caused by pooled water.

With these site preparation steps, you'll be ready to begin building your pool with confidence, knowing that your pool will be built on a solid, level foundation.

CHAPTER 5
STRUCTURE INSTALLATION

The pool frame is the support that holds the liner in place and provides resistance to water pressure. It is important to install the structure correctly to ensure the safety of the pool and to avoid future problems.

Here are the steps to install your pool frame:

1. Specification Check: Make sure you have the correct specifications for your pool structure, including depth, size and liner type. Make sure you have all the necessary materials and tools.

2. Frame Installation: Install the pool frame according to the manufacturer's instructions. This usually involves putting pipes or beams in place and filling with cement or other material to hold the structure together.

3. Alignment Check: Ensure frame is properly aligned and level. If necessary, adjust the structure before filling with cement or other material.

4. Installation of reinforcements: If necessary, install additional reinforcements in the pool structure to increase strength and safety.

These may include additional supports, reinforcing beams or other materials.

5. Final Check: Make sure the frame is completely installed and level. Make sure there are no loose parts or cracks in the structure before proceeding with siding installation.

With the structure installed, your pool will be ready to receive the liner and be filled with water. It is important to regularly check the structure of the pool to ensure that it is safe and free of problems.

CHAPTER 6
INSTALLATION OF THE COATING

After building the structural part, it's time to install the pool liner. The liner is the outer layer of the pool that provides protection and finish. Here are the steps to install your pool liner:

1. Choice of coating: Choose the type of coating that best suits your needs. There are several types of flooring available including vinyl, fiberglass, tile and more. Check the advantages and disadvantages of each type of flooring before making a decision.

2. Pool Preparation: Before installing the liner, make sure the pool is completely clean and free of debris. Remove any loose parts or bulges and level the pool surface.

3. Measuring and cutting the liner: Measure and cut the liner to perfectly fit the shape of the pool. Be sure to leave enough space for it to overlap the edge of the pool.

4. Liner Installation: Place the liner in the pool, starting at the bottom and working your way to the edge. Check that the siding is even and level and cut off any excess.

5. Fasten the liner: Secure the liner to the pool using clamps or other means of fastening, as indicated by the manufacturer. Make sure the coating is secure and has no bubbles or scratches. With the liner installed, your pool is ready to be filled with water and used. Be sure to follow the manufacturer's instructions for maintaining and cleaning the siding to ensure it lasts for years to come.

5. Fasten the liner: Secure the liner to the pool using clamps or other means of fastening, as indicated by the manufacturer. Make sure the coating is secure and has no bubbles or scratches.

With the liner installed, your pool is ready to be filled with water and used. Be sure to follow the manufacturer's instructions for maintaining and cleaning the siding to ensure it lasts for years to come.

CHAPTER 7

FILTERING SYSTEM

A filtration system is essential to keep pool water clean and crystal clear. It removes impurities and dirt from the water, keeping it safe and healthy for swimming. Here are the steps to install an efficient filtration system in your pool:

1. Choice of filtration system: There are several types of filtration systems available on the market, including pumps, filters and skimmers. Choose a system that suits your needs, based on your pool size and desired filtration level.

2. Pump Installation: Install the filtration system pump according to the manufacturer's instructions. This usually involves connecting the pump to the electrical power source and piping that takes water from the pool to the filtration system.

3. Filter Installation: Install the filtration system filter per the manufacturer's instructions. Check that all connections are tight and that water is flowing properly through the system.

4. Skimmer Installation: If using a skimmer, install it according to the manufacturer's instructions. The skimmer is responsible for removing the surface of the water, removing dirt and impurities before they reach the filter.

5. System Setup: Configure the filtration system so that water is returned to the pool at the proper pressure and flow. Check that the system is working properly, with no leaks or other problems.

With the filtration system up and running, your pool is ready for use. It's important to keep your filtration system clean and maintained regularly to ensure it's always working properly and keeping your pool water clean and safe for swimming.

CHAPTER 8
EQUIPMENT INSTALLATION

There are several optional equipment that you can install in your pool, each with its own specific function. Here are some of the most common fixtures and the steps to install them:

1. Ladder: Install a ladder in the pool for easy access to the water. Make sure the ladder is installed securely and stably in accordance with the manufacturer's instructions.

2. Lighting: Install underwater lights to light the pool at night. Choose bulbs that are appropriate for use in swimming pools and have a safety rating suitable for use in water.

3. Heater: If you want to use your pool year round, consider installing a heater. There are several types of heaters available, including electric, gas, or solar heaters. Install the heater in accordance with the manufacturer's instructions.

4. Cover: Install a cover on the pool to protect it from rain and excessive sun. This will help keep the water cleaner and prevent the energy needed to keep the water warm going to waste.

5. Chlorine System: If you are not using an automatic chlorine system, install a manual chlorine system to maintain it by adding chlorine to the pool water regularly. Follow the manufacturer's instructions to install the system correctly.

With all equipment installed and working, your pool will be ready for use. Make sure all equipment is maintained and cleaned regularly to ensure it is always working properly and safely. Always remember to follow all manufacturer's instructions and all applicable local regulations before installing any equipment in your pool.

CHAPTER 9
FILLING THE POOL

After all the previous steps have been completed, it's time to fill the pool with water. It is important to be patient with this step as it can take a few hours. Before you start filling the pool, make sure all equipment, including the filtration system, is working properly. In addition, it is important that the water used to fill the pool is adequate and free of impurities.

The first step is to turn on the water pump and wait until it is working properly. Then turn on the pool faucet and start filling the structure with water. It is important to constantly monitor the water level and adjust it if necessary to avoid damage to the structure.

After the pool is completely filled, you must wait at least 24 hours before using it. This allows the water to stabilize and proper chlorine and pH measurements to be taken. Remember to constantly maintain the water level, and to add chlorine and other chemicals as directed by the manufacturer.

Thus, you can enjoy your pool for a long time, safely and comfortably.

CHAPTER 10
CARE AND MAINTENANCE

Now that your pool is ready for use, it's important to follow a care and maintenance routine to keep it in good condition and ensure the safety of its users.

Pool cleaning is one of the most important precautions, and should be done regularly. It is recommended to use a sieve to remove leaves, insects and other debris from the surface of the water. In addition, it is important to empty the skimmer frequently and check the bottom of the pool for dirt.

Pool chlorination is another important care, and should be done frequently. It is important to follow the manufacturer's recommendations for how much chlorine to add to the water, and measure the chlorine level periodically.

Checking for leaks is another important activity. This can be done by monitoring the water level in the pool, which must be kept constant. If the water level drops, there may be a leak, and it's important to identify and fix it as soon as possible.

Keeping the filtering equipment and system in good condition is another important precaution. It is recommended to check the operation of the equipment regularly and carry out preventive maintenance when necessary. In addition, it is important to follow the manufacturer's recommendations for cleaning and maintaining the filtration system.

Remember that a well-maintained pool can last for many years, providing joy and fun for the whole family. Follow care and maintenance recommendations and get the most out of your pool.

CHAPTER 11
POOL SAFETY

Safety is an important concern when building a swimming pool at home. It is important to follow safety regulations and standards to ensure that the pool is safe for its users. Here are some important security measures to consider:

1. Pool Fencing: It is important to have a pool fence to prevent unauthorized access, especially for children and pets. Fencing must be at least 1.2 meters high and have gates that are closed and locked.

2. Ladders and Steps: Make sure your pool stairs and steps are secure and non-slip. This is important to avoid accidents.

3. Pool lighting: It's important to have good lighting around the pool to ensure visibility at night and to avoid accidents.

4. Alarms: Install pool alarms to alert if someone enters the pool unattended.

5. First aid: Have a first aid kit within reach of the pool and know how to use it in an emergency.

6. Rescue: Have rescue devices such as floats and lifebuoys within reach of the pool to ensure they can be used quickly in an emergency.

7. Training: Ensure all pool users have basic knowledge of swimming and pool safety.

Remember that safety is always the top priority when building a swimming pool at home. Follow safety regulations and standards, install safety measures and train your pool users on pool safety to ensure your pool is safe and fun for everyone.

CHAPTER 12
CONCLUSION

Congratulations! You have reached the end of this e-book on how to build a swimming pool at home. We hope you found the information you needed to get started on your pool building project.

Building a swimming pool at home is a major undertaking, but it can be quite rewarding. As well as being a great way to relax, a swimming pool can also be a valuable addition to your property.

However, it is important to remember that building a swimming pool requires planning and care, as well as regular maintenance to ensure that the pool is always safe and ready for use.

If you've followed the instructions in this ebook and followed the steps correctly, you're well on your way to having a home pool you can be proud of. Good luck with your project!

AUTHOR

Architect and urban planner since 2018, graduated from Centro Universitário Metodista – IPA, in Porto Alegre – RS. Postgraduate degree in Contemporary Education from the Instituto Federal Sul Riograndense in Charqueadas – RS.

Acting as a freelancer in the management and conduction of works and projects, since 2019 as an architect hired by the Municipality of Cachoeirinha - RS, coordinating the real estate cadastre and georeferencing sector. Also conducting works, such as the Pedreira Events Center in Eldorado do Sul, with more than 3000m² of built area implanted in a plot of more than 1 hectare, managing field teams and producing the various projects necessary for the development of the work.

Producer of digital manuals for civil construction, always aiming to provide a practical and easy-to-understand step-by-step, whether for investors or architects/engineers at the beginning of their careers.

Seeking to give the reader security in decision-making, clarity in processes and economy of time and resources.

HOW TO BUILD A SWIMMING POOL AT HOME

stay in touch

Instagram:
@rholmerphilipe

Email:
rholmercms@hotmail.com

Portfolio:
behance.net/rholmerphilipe

ISBN: 9798856567303

www.ingramcontent.com/pod-product-compliance
Lightning Source LLC
Chambersburg PA
CBHW072230290526
45794CB00007B/2963